C000143712

THE WORLD
ACCORDING TO
'NOT-THE-GURU' GAV

A concise beginner's guide to loving & healing
yourself, everyone & the planet Earth

GAVIN MUSCHAMP

BALBOA.PRESS
A DIVISION OF HAY HOUSE

Copyright © 2020 Gavin Muschamp.

All rights reserved. No part of this book may be used or reproduced by any means, graphic, electronic, or mechanical, including photocopying, recording, taping or by any information storage retrieval system without the written permission of the author except in the case of brief quotations embodied in critical articles and reviews.

Balboa Press books may be ordered through booksellers or by contacting:

Balboa Press
A Division of Hay House
1663 Liberty Drive
Bloomington, IN 47403
www.balboapress.co.uk
UK TFN: 0800 0148647 (Toll Free inside the UK)
UK Local: 02036 956325 (+44 20 3695 6325 from outside the UK)

Because of the dynamic nature of the Internet, any web addresses or links contained in this book may have changed since publication and may no longer be valid. The views expressed in this work are solely those of the author and do not necessarily reflect the views of the publisher, and the publisher hereby disclaims any responsibility for them.

Any people depicted in stock imagery provided by Getty Images are models, and such images are being used for illustrative purposes only.
Certain stock imagery © Getty Images.

ISBN: 978-1-9822-8233-2 (sc)
ISBN: 978-1-9822-8234-9 (e)

Print information available on the last page.

Balboa Press rev. date: 11/19/2020

Home Page.

To Mum & Dad
who have been with me through
thick and thin.
With much love and thanks.

Contents

Introduction.

I'm not a guru. I'm not enlightened. I'm not an expert. I'm still learning and I haven't got all the answers. So why would you want to read this book?

Well, during lockdown I have been avidly reading all kinds of inspiring books on self-help, self-empowerment and 'spirituality'. I feel really excited about it and coupled with the knowledge and experience I've gained over many years, I'd like to share some of this brilliant information with you.

So this is a brief, beginner's guide. A simplified, condensed version. Hopefully it will point you in the 'right' direction. There is a list, further on in the book, of the inspirational 'teachers' I've found so far, if you want to follow this further. And there are lots more to find!

You might not agree with it all, you might not like everything, but I'm asking you to keep an 'open' mind. Perhaps some bits will speak to you and other bits won't. That's fine! Maybe some bits won't make any sense now but later they'll make perfect sense!

I want to encourage you to think for yourself. Take self-responsibility. Find your own answers amongst the pages here, what resonates with your own 'highest' best self.

We're all going through big changes at the moment, whether we like it or not. So it's time now to look at ourselves, and to realise that we can all do something to improve ourselves and our situation and to move forward together towards a better world.

Gavin Muschamp
Sept. 2020

1 – In search of gurus and masters.

For 1000's of years the only way to get 'spiritual' knowledge was to search out gurus and masters in remote parts of the world. Now things have changed. There is a lot more information readily available to us, in books, CD's, DVD's and on the internet.

So maybe you thought I should look like this.....

But actually I look like this..... ⟶

2 – The old image of 'God'.

I used to think 'God' looked like this, (approximately).....

An old man in the sky

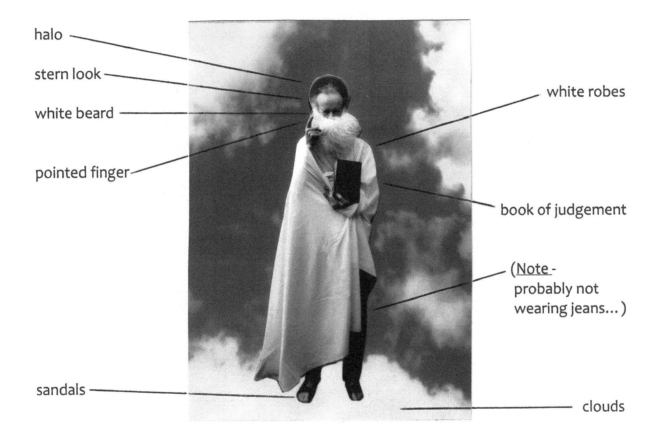

halo

stern look

white beard

pointed finger

white robes

book of judgement

(Note - probably not wearing jeans...)

sandals

clouds

Note – this photo was taken on a Sunday.

But now I think this..... →

Be your own Super Hero, (with God's help).

God wants us to express our own unique, 'best' individual self, part of Him/Her.

<u>**Note –**</u>
If you feel uncomfortable with the word 'God', feel free to substitute it for, Allah, Brahman, Jahweh, Akal Murat, Divine Being, the Universe, the Source, Spirit, Life Force, the Light, Unconditional Love, Higher Self etc or whatever feels right for you.

3 – Love your whole self.

- Look after yourself physically, emotionally, mentally and spiritually.
- Remain in balance.

- Some suggestions – Exercise, walks in beautiful Nature, eat a healthy diet, eat organic, yoga, meditation, keep a journal, read inspirational books, do things you love and feel passionate about.

Lea Rhododendron Gardens, Derbyshire.

4 – Find some quiet and stillness.

- Amidst the busy world. Slow down for a while.
- Be in the present moment.
- Experience peace inside & outside.
- Have some time on your own.
- Walks in Nature.
- Meditation.
- Listen to the quiet voice within, (intuition, Higher Self and 'God').

5 – Life is a school and an adventure.

- Possible problems, challenges, difficulties and 'suffering' come up for us all in our lives. They can affect every one of us. But it's not what happens to you, but how you react to what happens to you that matters.
- See problems as opportunities, (to learn, improve and grow).
  Suddenly they don't seem so bad!
- Think of life happening <u>for</u> you, not to you.
- Face your fears.

6 – Be Positive!

- Be optimistic.
- Look on the bright side.
- Find ways to be happy.
- Smile!
- Use affirmations.

T-shirt from 'Chargrilled'.

Wirksworth Market Place, Derbyshire.

7 – Gratitude.

- We have so much in life already!
- Be thankful for the abundance you already have.
- Don't take things for granted.
- Be generous. (Give as well as receive).

Life <u>is</u> a bowl of cherries!

8 – Manifest your own 'reality'.

- Follow your dreams.
- What you think about and focus upon you will get.
- Like attracts like.
- Imagine the highest good for yourself and everyone.

South Bank, London.

9 – Be Yourself!

- Take self-responsibility.
- Move from victim to self-empowerment.
- Don't just 'fit-in'. (Don't be afraid to stand out).
- Make up your own rules. (But don't harm others).
- Express your own unique, individual self.
- Find like-minded souls.

Greenhill, Wirksworth, Derbyshire.

T-shirt from 'Chargrilled'.

10 – Find your purpose.

- Do what you love.
- Let your heart sing.
- Don't hide your light away.
- Let your light shine!

T-shirt from 'SubtleSplit'.

Church Walk, Wirksworth, Derbyshire.

11 – Use your full potential.

- Use all of your abilities, skills and passions.
- Get excited about life! (Don't get bored).
- Don't play small with your talent.

Findhorn Foundation Community, Moray, Scotland.

12 – You don't have to be 'perfect'.

- Don't try and please everyone else all the time, (it's impossible!).
- Accept yourself exactly as you are, (warts and all!).
- Accept your 'shadow' side, don't deny it. Bring it into the light of love.
- Doing your 'best' is 'good enough'.
- Risk sometimes. Don't be afraid to get things 'wrong'. We often learn a lot from our 'mistakes'.

Findhorn Foundation Community, Moray, Scotland.

13 – Gain knowledge, wisdom and understanding.

- Be inspired. Be inspiring.
- Talk to exciting people.
- Learn from others.
- Better yourself and your understanding.

Cromford, Derbyshire.

14 – Keep an 'open' mind.

- Think all possibilities.
- Don't limit yourself, (and put yourself in a box).
- Be flexible.
- Step into the unknown.
- Challenge yourself.
- Get out of your comfort zone.
- Think, 'I can', rather than, 'I can't'.
- Try something new and surprise yourself.

15 – Have fun, play and laugh.

- Make time to relax and enjoy yourself.
- Do fun things with your family.
- Laugh with your friends.
- Be creative.

T-shirt from 'Chargrilled'.

16 – Don't take yourself too seriously.

- Be a bit detached.
- Be objective.
- Having a sense of humour helps you to see things in perspective.

17 – Simplify, clarify and de-clutter.

- Helps you think more clearly and feel free.
- Let go of the past, (anything that is holding you back).
- What is important in life?
- Make your environment beautiful.

Wirksworth recycling area.

18 – Go with the flow.

- Everything happens for a reason. (It might not be obvious why until later!)
- Dance with life.
- Don't try and be in control all the time.
- Let go and see where it leads you, (but don't be irresponsible).

T-shirt from 'Chargrilled'.   Matlock Bath, Derbyshire.

19 – Accept everyone!

- We're all equal.
- No-one is better than anyone else.
- Let go of judgement and criticism.
- Have compassion.
- Accept our differences, we're all unique.
- Accept everyone whatever colour, nationality, religion, (or no religion), class, size, shape, hair style, (or no hair!) and perceived 'weirdness'!

20 – Serve others.

- Help everyone.
- Work together.
- Co-operate, (a win-win situation).
- It's not just about you!
- Give something back to your community.
- Share your skills.

NHS, (National Health Service) UK.

21 – We're all in this together.

- We all share in common the experience of being Human.
- We're all one!
- Treat others as you would like to be treated yourself.

T-shirt from 'Chargrilled'. St. Mary's Church, Wirksworth, Derbyshire.

22 – We have a choice now.

- We have the opportunity to change things for the better.
- To learn to love and heal ourselves, everyone and the planet.
- Be the change you would like to see in the world.
- Look after the Earth and all the life forms we share it with. This is our beautiful 'home'.

Matlock Youth4Climate strike.
Matlock, Derbyshire.

23 – Why wait?

- Do it now!
- Before it's too late.
- Life is short! (Have no regrets).
- I thought growing old would take longer!

Conclusion.

This book is only a brief, beginner's guide. To point you in the 'right' direction.

To find out more, there are lots of inspiration 'teachers' out there right now, for you to look into and explore.

Whatever 'speaks' to you. They are all saying the same thing, but in different ways, from different perspectives. To appeal to different people.

As well as learning a great deal from these 'teachers', ultimately you have all the answers for your 'best', unique you within yourself. So I want to encourage you to find your own answers also, by listening to your own intuition, (inner teacher) or 'Higher Self'. (Your direct connection with 'God').

You are a unique individual Human Being. You do count, whoever you are, wherever you are. We are all an important piece of the jigsaw, contributing to the wholeness of life. (Universe is the unity of diversity). We are all learning to work together to love and heal ourselves, everyone and the planet Earth itself.

So here is one last thought.....

The main message to remember is.....   ## Move from fear to LOVE.
Focus on love, think love, be love.

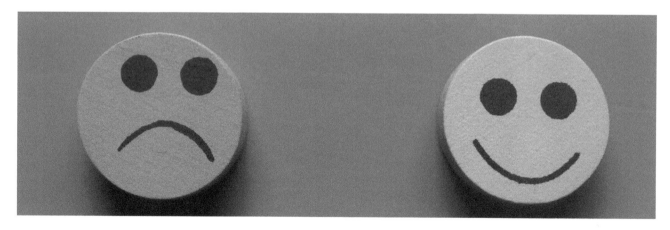

Blessings!

Not-the-Guru Gav, (Gavin Muschamp).

Acknowledgements.

Here are some of the many diverse people I've been inspired by and there are many more to find! (In alphabetical order).

Amma (Mata Amritanandamayi).
Sir David Attenborough
Richard Bach
Melanie Beckler
Rhonda Byrne
Eileen Caddy
Diana Cooper
Patricia Cota-Robles
Ram Dass
Mike Dooley
Dr Wayne Dyer
Jeff Foster
Louise Hay
Susan Jeffers
Vex King
Robin Sharma
Haemin Sunim
Greta Thunberg
Eckhart Tolle
Tim Whild
Stuart Wilde
Oprah Winfrey
Maharishi Mahesh Yogi

Appendix.

Disclaimer.

I'm afraid there are no guarantees following the guidelines in this book, (although I sincerely hope they will help).
It is not intended as a substitute for seeing a doctor or mental health expert. If you are having health problems, (especially mental ones), it is recommended that you seek the relevant medical advice.

About the author.

Gavin Muschamp is interested in self-help, self-empowerment, spirituality and green issues. He has been practising meditation for many years and has read widely on spiritual subjects. He spent 6 years living in spiritual communities.

He also has a BA Visual Arts and is an artist and writer who likes dressing-up, (but up until this point hadn't found an outlet for this secret passion!).

He lives in Wirksworth, a small quirky, characterful town on the edge of the Peak District National Park in Derbyshire, England. It is home to many artists and creative people. (There is an Arts Festival every year in September).

T-shirt from 'Chargrilled'.

<u>Note</u> – If you are interested in seeing my paintings and eco-friendly greetings cards, please visit, **www.gavinmuschamp.com**

Lightning Source UK Ltd.
Milton Keynes UK
UKHW051026271120
374134UK00006B/24

9 781982 282332